"One who
plants a
garden
plants
happiness."

Camille Soulayrol

Photography by
Frédéric Baron-Morin

Plantopia

CULTIVATE · CREATE
SOOTHE · NOURISH

Flammarion

Contents

Green is the new black   8

1. Caring for Plants at Home: 12
    An Introduction

DEVELOP YOUR GREEN THUMB   14
Basic advice for absolute beginners   16
No-stress plants for the novice   18
Repotting   26
Taking cuttings   28
DIY • Propagating a Swiss cheese plant   30
DIY • Propagating a *Pilea*   32

PURIFYING PLANTS   34
Plants to clean the air   35
Eight air-purifying plants   36

CACTI AND SUCCULENTS   42
What you need to know   44
Three golden rules   45
Three mistakes to avoid   45
Starter cacti   46
Foolproof succulents   49

2. Plants for a Beautiful Home   52

TERRARIUMS   54
DIY • Closed terrarium   56
DIY • Open terrarium   58

FLOWERS AND FOLIAGE   60
DIY • Drying plants   62
DIY • Framing pressed leaves   64
DIY • Wall decorations   66
DIY • Dried flowers   68
DIY • Dried-flower wreaths   70
DIY • Foliage wreath   74
DIY • Leaf-trimmed hoop   78

Inspiration • Touches of green   80

PLANTERS   82
DIY • Upcycled pallet garden   82
DIY • Standing planter   84
DIY • Hanging planter   86
DIY • Trellis of triangles   88
DIY • Himmeli   90
DIY • Hanging bottle planters   92

MACRAMÉ   94
DIY • Basic macramé knots   94
DIY • Hanging planter for beginners   98
DIY • Advanced hanging planter   100
DIY • Macramé planter stand   102

Inspiration • Creative containers   104
Inspiration • Fun with foliage   108
Inspiration • Aquatic plants   110

3. Going Green without Chlorophyll   112

A GREEN HOME   114
DIY • Plant-based dyes   118
DIY • Botanical illustrations   122
DIY • Embroidery for beginners   124
DIY • Advanced embroidery   128
DIY • Paper jungle   132

4. Plants for Well-Being   136

ESSENTIAL OILS AND BEAUTY DIY   138
What you need to know   138
Must-have oils   140
DIY • Essential-oil diffusers   144
DIY • Lip balm   146
DIY • Hydrating body lotion   147

INFUSIONS AND RECIPES   148
What you need to know   148
Must-have infusions   150
DIY • Infusions for well-being   152
DIY • Waters and "green juice"   154
DIY • Eat your greens!   156

Acknowledgments   158

Green
is the new black

Plant stores seem to be sprouting up left and right these days, some specializing in succulents and purifying ficus trees, others crammed with dried leaves and flowers for DIY decoration. Friends get together on the weekend to swap plant cuttings, and a *Pilea* plant now makes a much better hostess gift than a bouquet. Bearded men deliver flowers by bike, and online subscription services deliver seeds right to your doorstep. Grass juices have replaced multi-vitamins, flowers garnish salads, and herbal teas are the new cocktails. Enter any concept store in search of fashionable clothes and you're likely to leave with a cactus in hand or a macramé trinket crafted by the designer herself. Lush motifs climb our walls and our clothes, and flower garlands are flourishing on social media. Nature is everywhere!

Sure, it may be a fad, and some people may have had their fill already. But this return to nature is actually a very good sign, because plants remind us of what really matters: **that life is precious and that nature needs us—but not nearly as much as we need it.** If you know how to listen, nature will gladly share with you its many benefits for mind and body.

You don't have a green thumb? Plants die as soon as you touch them, though you'd like to give them a try? Then this book is for you. Trust me, I adopted a *Pilea* at the beginning of the project—I mother it like a baby— and I started a collection of dried plant specimens that seems to grow by the day.

In other words, no more excuses. It's time to plant your first seed.

Welcome
to the
green life!

1. Caring for Plants at Home: An Introduction

Develop Your Green Thumb

In the following pages, Mama Petula, the high priestess of the urban jungle, shares her top tips and advice for choosing and caring for your "soul plant." Caroline Ciepielwski, owner of Mama Petula, is a pioneer in green living. After working for others for years, this professionally trained gardener started taking on independent plant decoration projects. Eventually, she found a wonderful space in Paris's urban art collective Grands Voisins and opened her first store, an oasis where she looks after plants as if they were her own children. Coaxing beauty from next to nothing, letting nature take control, and sharing a love of living things is the spirit behind Mama Petula.

Basic advice for absolute beginners

Rule No. 1 // Observation

Plants are just like us: living things that need water, food, light, and companionship. Sound familiar? Good. If you respect their needs, plants will become your best friends, each with its own personality. Get to know them, watch them grow, listen to them, and they'll let you know what they need. Pay attention—they have a lot to teach you.

Rule No. 2 // Light

You'll often hear it said that plants need light. This is true, but be careful. Light doesn't necessarily mean sunlight. You probably find sipping a coffee in the sun wonderful, but how does your skin feel after a long day at the beach? Depending on the plant and its needs, find it a light-filled or shady spot, but avoid direct sunlight. Houseplants are exotic plants used to living in the jungle where little sunlight penetrates the canopy. See? It's not rocket science.

Rule No. 3 // Watering: a prickly subject

Accidents often arise from our best intentions. The fact is, most people drown their plants. It's impossible to tell you precisely how much water to give your plants, because that depends on things like conditions in your house, where the plant is placed, and its size. But we'll give you some advice and recommendations. Remember, there's no replacement for interaction and observation.

TIP Water your plants more abundantly during the growing season (spring, summer, and fall) than in periods of vegetative rest (winter). Don't use cold water: room-temperature is better.

Rule No. 4 // Repotting

Every healthy plant grows and needs more and more space, just like you. You'll know when the time has come by looking at its roots: if you start to see them curling around the soil in a ball, the plant is suffocating. You should repot it in a pot at least 3/4 in. (2 cm) larger, fill in with good-quality soil (rich in nutrients) or add fertilizer if the soil is too poor, and let it grow.

Rule 5 // The container

Plastic pots should be banished, especially those that don't allow water to drain away. There's nothing worse for a plant than stagnant water. Use earthen pots instead.

Rule 6 // Attention

Plants need love, friendship, companionship—whatever you want to call it. It might sound crazy, but I know it's true. "Anyone with plants will tell you the same thing," says Mama Petula. They need care and attention. Don't abandon them when you go on vacation, do talk to them when you get home from work, and look after them tenderly.

To own a plant is to show a healthy respect for living things.

No-stress plants for the novice

Calathea
Easy-going

Common name: Peacock plant
Measuring around 8 in. (20 cm) tall, this strikingly patterned
plant grows quickly and tends to expand out, rather than up.
Bonus: It prefers shade to light and likes moist environments
and regular watering.

Begonia maculata
Generous

Common name: Polka-dot begonia
This plant will make you the queen of cuttings, enough
to astound even your most green-fingered of relations.
This flowering plant with patterned leaves grows before
your very eyes and prefers shade to light.
Bonus: It grows rapidly but also gets really thirsty.
So be vigilant and remember to water it regularly.

Fittonia

Striking

Common name: Nerve plant

The eye-catching veins on this ground-cover plant look like they were drawn with a felt-tipped marker. Mama Petula nicknames it "the pouf plant" because it grows into a shape resembling a comfortable beanbag. It likes gentle light and doesn't do well in direct sun. It needs spare but regular watering.

Bonus: Don't worry, it will let you know when it's thirsty: it wilts, but perks up as soon as you water it.

Pilea peperomioides
The hipster

Common name: Chinese money plant
Pilea is considered a lucky charm in China because
of its coin-shaped leaves. According to legend, it will
bring prosperity if you bury a coin in the soil (I'll try it
and let you know). It likes sunlight and needs regular
watering, so check the soil once or twice a week.
Bonus: It's a great plant for taking cuttings.

Pteris faurei

Thirsty

Common name: Fern

This plant is for anyone who's a little heavy-handed with
the watering can. Sound familiar? It does to me. Ferns belong
to the group of plants that make up forest undergrowth.
As such, they need little light and a lot of water. Be careful
to respect their love of shade and moisture. They would
be very happy in a terrarium.

Bonus: Ferns do well in bathrooms and basements.

Monstera deliciosa
Extroverted

Common name: Swiss cheese plant, monster fruit, split-leaf philodendron

Monstera deliciosa is a bold and energetic climbing plant that grows as vigorously inside as it does in the jungle, giving an exotic touch to interior spaces. Mama Petula lovingly refers to it as "jungle weed" because it loves life and grows rapidly pretty much anywhere. It has quite an appetite and needs an abundance of nutrients, water, and light.

Bonus: You can take many cuttings for friends and family (see page 28).

Aeschynanthus
The flowering hanging plant

Common name: Lipstick plant
Measuring about 16 in. (40 cm) in length, this plant needs
light but not direct sun. Check every ten days to see if it needs
water. If it starts to lose its leaves, it's high time to water.
Bonus: The flowers are dazzling.

Senecio rowleyanus
The "hip" hanging plant

Common name: String of pearls
This forgiving trailing plant will tolerate
forgetfulness, so give it a try. This succulent
grows to around 12 in. (30 cm). As
a hanging plant, it will do well on
the corner of a dresser or in a homemade
macramé hanger (see page 98). It needs
light (remember, light doesn't mean direct
sunlight) and moderate watering; in other
words, about once every two weeks. But
nothing can replace getting your hands
in the soil and giving it regular attention.
Bonus: This is the perfect plant for
getting started—it's very tolerant
of forgetful beginners.

TIP Hanging plants are often located
in high places where heat rises, so they
will probably need more water than
desk-top plants.

2. Add a layer of clay pebbles at the bottom of the pot to keep the roots from blocking the hole. This will help good drainage.

1. Find an old, damaged pot. Wrap it in a towel and break it into small pieces using a hammer, then place a few crocks at the bottom of your new pot. This will prevent soil from blocking the drainage hole(s).

3. Choose a pot 3/4–1¼ in. (2–3 cm) larger than your plant.

4. Remember to give it a treat: finish with a little fertilizer (plant food).

Repotting

Plants grow, just like us. Remember how uncomfortable you felt as a growing kid in too-tight clothes? Give plants more space by transferring them to larger pots, ideally 3/4–1¼ in. (2–3 cm) larger. You should do this every two to three years. Really robust plants may need to be repotted more often.

Plants on sale are often cultivated in poor conditions, so be careful when making your purchase. You may find they've been cramped and were waiting for you to come along and free them.

·Determining if a plant needs to be repotted is very simple: it will let you know. Remove the pot and look at the roots. If they are wrapped around the soil, as illustrated in the image opposite, it means they don't have enough space. They are knotting up and may suffocate the plant. It's time to act. It's also a good time to add compost and nourish the plant with a little fertilizer. Finish with a good watering and your plant is set to grow.

TIP Roots, like vampires, love to hide in soil and hate the light. Keep this in mind when repotting and avoid extended exposure to the sun.

Taking cuttings

This method for propagating plants is a cinch. Give plant cuttings
a try and watch your house transform into a wizard's workshop.
The results can be quite magical.

The secret: Cuttings are most successful when taken during
a growing season, especially in spring.

1. Cut the stem on the diagonal with a clean tool (knife, scissors, or shears). Always remove the cutting above a node on the mother plant, then cut once more below a node on the cutting itself.

2. Remove any leaves from the base of the stem. There should be no leaves below the water line.

3. Place the stem in a vase or a bottle and add a little water. Let the cutting soak and watch the roots develop. Change the water if necessary.

4. When the roots are over 1¼ in. (3 cm) long, you can transfer the plant to a pot.

aerial
roots

Propagating a Swiss cheese plant

Monstera fans, this is for you! These generous plants are easy to propagate through cuttings. This quick lesson will have you planting them all over your house and treating your friends to some plants of their own.

Find a branch with a short aerial root growing from it. As before, cut above a node and put the stem in a transparent vase so that you can watch the roots develop. When the roots are 1¼–2 in. (3–5 cm) long, you're ready to repot: you've got yourself a new plant.

You can also just leave it in a transparent jar and watch the long roots grow.

This attractive houseplant is also easy
to propagate through cuttings: you'll make
a lot of people happy. And remember,
it brings good luck.

Propagating a *Pilea*

roots

plantlet

1. Using your fingers, separate the plantlet (young plant) from the mother plant without removing it from the soil.

2. Using a small knife, cut vertically into the compost, making sure you include several of the plantlet's roots. Cut deeply and cleanly.

3. Once the plantlet is removed, make sure there are roots attached so you can repot it.

4. Place clay pebbles and soil in a container (as described on p. 26) and repot the plantlet. Water it well. Slip a coin in the soil and get ready to meet Lady Luck.

Purifying Plants

Raised among flowers and trained by plant enthusiasts, Agnès Valverde, owner of the Parisian flower shop Éphémère, helps us breathe more easily by sharing her expertise about plants with purifying powers. For the last twelve years, this great-granddaughter of arborists who specialized in the cultivation of peonies has run a lovely little boutique where she sells high-quality flowers and plants chosen with love.

Plants to clean the air

Help! Our houses are toxic, and unidentified poisons are hiding everywhere, in cleaning products, paint, building materials, and who knows where else.

But don't panic. You don't have to drop everything and flee to a desert island. Plants are incredible natural remedies. Night and day, plants absorb and filter air pollutants through their leaves, then break them down in their roots, which convert them into perfectly safe organic compounds consumed by the plant.

Through the process of transpiration, plants emit oxygen in the form of water vapor, which renews the atmosphere. And every plant does this.

Agnès Valverde has chosen several easy-going no-stress varieties that will have you breathing more easily.

P.S. A plant's purifying power is connected to its rate of growth. The faster a plant develops, the more air it uses, and the more vigorously it will filter pollutants.

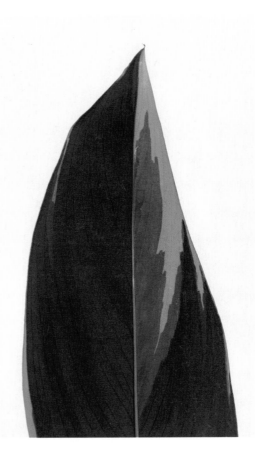

Eight air-purifying plants

Hoya carnosa 'Tricolor'
The creeper

Common name: Wax flower
This decorative cascading plant produces clusters
of star-shaped flowers.
Bonus: Its exotic appearance will transform
your living room into a jungle.

Tradescantia 'Yellow Hill'
The easy trailing plant

Common name: Inch plant
This low-maintenance trailing plant tolerates medium
light but can also flower in full sun.
Bonus: It is the queen of cuttings. Cut 4 in. (10 cm) stems
and keep them in water until roots appear. Repot them
in soil specifically created for potted plants.

Sansevieria cylindrica
The self-sufficient succulent

Common name: African spear, cylindrical snake plant
You can take a vacation worry-free with this plant around:
it has to dry completely between waterings and can
go several weeks without water. It's perfect for the absent-
minded and for world travelers. It also likes a lot of light.
Bonus: It has a funky jungle look.

Chlorophytum comosum 'Variegatum'
The no-brainer

Common name: Spider plant
This one grows fast and needs a lot of water. Its prolific roots
develop quickly and consume great quantities of air,
making it an incredible air filter.
Bonus: It doesn't need much light.

Scindapsus aureus
The fast-grower

Common name: Devil's ivy
This generous, climbing creeper doesn't need much upkeep
and will quickly take over your home. It likes a lot of light.
Bonus: It grows quickly, and its expansive personality
combines beauty and purifying power.

Spathiphyllum
The beauty queen

Common name: Peace lily
This easy-going plant flowers yearly between
May and October.
Bonus: It's communicative, which beginners will appreciate.
Its leaves droop when it's thirsty, a sign it needs watering.
The perfect first plant.

Cacti and Succulents

Le Cactus Club is a Parisian shop run by sisters Pauline and Camille, who specialize in plant decorating, cacti, and succulents. Raised by a father who loves gardening and a mother with a passion for interior design, they decided to create a space that combines the two. Here they give their expert advice for choosing the best plants and keeping them happy.

What you need to know

Understanding the difference between a cactus and a succulent

The term "succulent," from the Latin *succus*, means "full of juice." It describes the ability of certain plants to stock a large quantity of water in their leaves, stems, or roots, enabling them to survive long periods of drought. So it's not a family of plants per se, but rather a capacity shared by plant varieties from a number of different species, including cacti.

All cacti are succulents, but not all succulents are cacti.

Easy-going plants

Cacti and succulents are easy-to-maintain houseplants because they like our warm, dry homes. Originally from arid regions, they rarely need to be watered and are perfect companions for the absent-minded. What they really need is good light. If you want to see them flourish, keep them in the brightest areas of a room, near the windows.

Three golden rules

Rule No. 1 // Water
Remember, it rarely rains in a cactus's natural habitat, but when it does, it's torrential. Succulents need a good soaking, but only about once or twice a month. Adjust the frequency of watering according to the season and the size of the pot, and let the soil dry completely between each watering.

Rule No. 2 // Light
Cacti may not need much attention, but they do crave light. These desert natives are constantly exposed to direct sunlight and only a few varieties tolerate shade. At home, they are best placed on a windowsill, on a sunporch, or under a skylight.

Rule No. 3 // Repotting
Plan on repotting your cacti and succulents every two to three years, as you would any other houseplant, in a pot larger than the last, and ideally in spring or summer. Use soil specially formulated for cacti/succulents/bonsai, which is particularly light, and add a layer of clay pebbles to the bottom of the pot to let excess moisture drain away.

Three mistakes to avoid

Mistake No. 1 // Stagnant water
Succulents detest stagnant water. It prevents the soil from drying out and encourages bacterial growth, which leads to rot. Pots must have holes to enable drainage. If you'd like to use a decorative planter without drainage holes, keep the plant in a plastic pot with holes so that you can remove it for watering.

Mistake No. 2 // Overwatering and humid rooms
While it's pretty easy to correct for under-watering, overwatering is often fatal. If you can't remember the last time you watered your cactus, err on the side of caution and wait a week or two longer. Choose well-ventilated areas in your home, and avoid putting succulents in the kitchen or bathroom, which are often too humid.

Mistake No. 3 // Winter on the balcony and vacation in the dark
Cacti and succulents are saturated with water, so they are unable to withstand freezing temperatures. They are happy to stay outdoors in good weather, but keep them indoors from early fall to spring. When you leave on vacation, water them before you go and remember to leave the drapes open to let in some light.

Starter cacti

Opuntia ficus-indica
The edible cactus

Common name: Barbary fig, prickly pear
This variety, endowed with many flat pads, is an important ingredient in Mexican cuisine. While the plant is particularly known for its fruit, its pads are also edible once their spines have been removed.
Bonus: The Barbary fig is a superfood, with many medicinal properties.

Myrtillocactus geometrizans
The movie star

Common name: Blueberry cactus
Found most commonly in Mexico and Guatemala, this cactus is seen on the set of every Western film and is the one that fans cultivate most widely. In nature, it can reach up to 16½ ft. (5 m) tall and 8 in. (20 cm) in diameter, forming spectacular cactus forests.
Bonus: It has few spines, making it perfect for households with children.

Mammillaria spinosissima
The cactus for beginners

Common name: Spiny pincushion cactus
Mammillaria is a small, easy-growing cactus found in South America and the West Indies. It gets its name from the rows of nipple-like nubs that cover its surface. Some varieties, known as "fishhook cacti," have curved spines that reel in the unwary.
Bonus: Of all the cacti, this one flowers indoors with the least difficulty. It produces an adorable crown of pink flowers at its apex.

Epiphyllum anguliger
The tropical cactus

Common name: Fishbone cactus, zigzag cactus
Those notched leaves make it hard to believe this one's a cactus. Unlike most cacti, it comes from humid tropical forests and needs regular watering and consistent moisture. The incredible flowers can reach 8 in. (20 cm) in diameter.
Bonus: It's one of the few cacti that tolerate partial shade.

Echinocactus grusonii
The orb

Common name: Golden ball, mother-in-law's cushion
This is one of the most popular cacti thanks to its unique and exceptionally even shape. Long threatened with extinction from illegal harvesting, it is now a protected species. It is one of the most tolerant to drought: you can forget it for literally months and it will be fine.
Bonus: It grows slowly, which makes it perfect for small spaces.

Foolproof succulents

Kalanchoe tomentosa
The teddy bear

Common name: Panda plant
This succulent, native to Africa, has particularly elegant foliage. Its long leaves are trimmed in red and covered with a thick, silvery velvet. Placing it in bright light will help maintain the color contrast and peachskin fuzz.
Bonus: The *Kalanchoe* family of plants includes a large range of shapes and colors.

Ceropegia woodii
The romantic

Common name: String of hearts
The poetic appearance of this South African plant has made it quite fashionable. Its fine stems lined with heart-shaped leaves can grow up to 8 ft. (2.5 m) long. A trailing plant with little bulk, it looks lovely on a bookshelf.
Bonus: You can prune it to the desired length at any time.

Echeveria

The desert flower

With leaves that form a rosette, these succulents
resemble flowers. It's better to water plants with compact
rosettes by capillary action—soak them as opposed to
top-watering—to avoid water collecting between the leaves.
Bonus: Propagation is a breeze. Simply break off a leaf
and stick it in soil to get a new plant.

Euphorbia erythraea
The imposter cactus

Common name: Candelabra tree
This succulent doesn't actually belong
to the cactus family—it's a Euphorbia.
Its spines are actually fleshy growths,
and it contains a milky sap or latex.
Careful, it's toxic. Wash your hands
after handling the plant.
Bonus: It grows much faster than cacti.

Crassula ovata
The lucky charm

Common name: Jade plant
This classic succulent is an incredibly
popular little tree. Very easy to maintain,
it's also the perfect gift—supposedly,
it brings good luck and prosperity. You can
let it grow freely or prune it like a bonsai.
Bonus: Certain varieties of Crassula take
on a lovely red hue when they are exposed
to the sun.

2. Plants for a Beautiful Home

"Plants are no longer objects of desire, they are necessities."

—Noam Levy

Terrariums

Noam Levy, the owner of Green Factory, has helped popularize terrariums in France and beyond. He's been experimenting with these "mini worlds" for over ten years, exploring the possibilities they provide for greening up our homes and lives. His creations illustrate life lived in harmony with nature and show us how we might live together intelligently.

While we may dream of living surrounded by plants, their needs are often very different from ours. The terrarium technique enables humans and plants to thrive under the same roof. Living together in harmony, each in its own perfect ecosystem: if only such a thing existed for couples …

What is a terrarium?

Terrariums are small-scale reproductions of the life cycle, from photosynthesis to the water cycle. They're little miracles, really. Having a terrarium at home is to be reminded of life's essentials on a daily basis.

Key advice A terrarium is a small, self-sufficient world. It doesn't need to be watered regularly; one or two waterings a year should suffice. It needs light, but not direct sun.

Golden rule

For a successful terrarium, choose plants that will flourish in this environment. Don't hesitate to consult a specialist before you get started.

We've put together two DIY methods for creating your own terrarium: the **closed terrarium**, designed to preserve the water cycle and condensation formation; and the **open terrarium**. The plants in each environment are, of course, very different, so choose the one you prefer. But remember, nothing's stopping you from making one of each.

DIY Closed terrarium

Supplies
- green plants that like moisture (such as fern, creeping fig, dwarf begonias, or ivy)
- glass jar
- pumice or clay pebbles
- fine gravel, in different colors if desired
- soil mixed with a little coarse sand
- moss
- decorative rocks and stones

1. Line the bottom of a glass jar with a 1¼ in. (3 cm) layer of pumice. Add the gravel in several layers to create an attractive visual contrast.

2. Pour a ¾–1¼ in. (2–3 cm) layer of soil on the pumice and make a small well for your plants.

3. Slightly loosen the plants' root balls by carefully squeezing them in your hands. Place them in the well and press firmly so that they sink an inch or so into the soil, down to the pumice.

4. Place the moss around the plants, green side up, and arrange the rest of the rocks and stones decoratively. Be careful not to pack the soil too tightly.

5. Clean the interior and exterior of the container. Pour water down the walls using a quick, circular motion, and lightly water the soil surface.

TIPS A wet terrarium should always remain closed, except when temperatures spike. The temperature inside a glass container is on average 20° F (10° C) hotter than room temperature. In case of a heat wave, open the terrarium to lower the temperature. To determine if the plant is too hot, check to see if large drops are dripping down the jar's interior. If they are, it's sweating too much. However, a little condensation is perfectly normal. Once it's been planted, never water a terrarium, but do spray it once a year with distilled water. To clean the container, use water and a little paper towel and perhaps a bit of white vinegar if it's really grimy. Never place a terrarium in direct sun, but do give it lots of light.

TIP To encourage healthy growth and beautiful blooms in cacti, respect their winter sleeping period. This means giving them enough light, and keeping them cool (between 40–50° F, or 5–10° C), especially at night.

Open terrarium

Supplies
- cacti
- vase or wide bowl
- pumice or clay pebbles
- colorful gravel
- soil mixed with a little rough-grain river sand
- decorative stones and pebbles

Dry terrariums are designed to grow cacti and should remain open. They can be placed in the sun, since these plants love light. Native to dry habitats, cacti adapt by stocking water in their leaves, stems, and roots. Avoid spraying the leaves and water the soil sparingly, at most once a month in winter and twice a month in summer.

1. Line the bottom of the vase or bowl with ¾–1¼ in. (2–3 cm) of pumice or pebbles. Add the gravel in several layers to create pleasing contrasts.

2. Pour the mix of soil and rough-grain sand on the pumice and make a small hole for your cacti.

3. Gently loosen the root balls by carefully squeezing them between your hands. Plant the cacti in the well and press down firmly, so that they sink an inch or so into the soil, down to the pumice.

4. Arrange the stones and remaining gravel to produce a decorative effect. Clean the container and lightly water the landscape, being careful to avoid the cacti.

ATHYRIUM FILIX-FEMINA

Woodsiaceae · Fougère Femelle · Hémisphère Nord

FORTUNE ET PROTECTION

Flowers and Foliage

We remember collecting and pressing flowers as children but tend to forget about it when we grow older. This simple pleasure is well worth rediscovering. When I started this project, I had a blast putting together a collection of dried, pressed botanical specimens. It's so easy and so rewarding. And what fun it is to turn afternoon walks into treasure hunts. A forest fern, some dried baby's-breath in a bouquet: nature invites itself into our homes even in the dead of winter. This is not a lesson from an expert, but I hope it will inspire you to explore further. Perhaps you'll even go on to become a famous naturalist.

 DIY Drying plants

The secret is to dry the flowers and leaves in newspaper (be careful not to use glossy paper). The daily paper will do just fine—plus you'll be supporting the press and getting a good read into the bargain.

Just remember to always take a pair of pruning shears and a little plastic bag with you when you go out walking.

1. Carefully spread each cutting on a sheet of newspaper and place another sheet on top. Feel free to use a little tape to keep each one nice and flat. Be careful not to use wet cuttings so as to avoid mold.

2. Slip the prepared cuttings between the pages of a book and place something heavy on top to maintain an even pressure.

3. Let dry between one and three weeks, taking a peek from time to time. You can also change out the newspaper, which will absorb moisture from the plant, and see how things are progressing.

TIP Obviously, some plants are easier to press than others, but it's still quite fun to try different things and see how they turn out. You'll likely get some nice surprises.

Supplies
- flowers and leaves
- newspaper
- craft tape
- large book and heavy weight

Framing pressed leaves

Supplies

- dried flowers and leaves
- two panes of glass of equal size
- decorative tape or washi tape

Once the flowers and leaves are dry, have the panes of glass cut to size at your local hardware store. Slip them between the panes of glass and seal the edges with tape to add color and create a decorative effect. You can also use all-glass picture frames in place of panes.

 DIY | Wall decorations

Dried flowers and leaves are far too pretty to stay hidden away between the pages of a book. Here's a simple way to enjoy them.

Supplies
- dried flowers and leaves
- pages from an old book
- paper of various textures
- washi tape
- sticky tac or double-sided tape

1. Remove the pages from the book. Choose a variety of paper in different sizes and textures to make a wall-sized gallery, then arrange the flowers and leaves on the pages.

2. Attach the plants to the paper using little strips of washi tape, then hang them on the wall with a bit of sticky tac or double-sided tape.

And there you have it: a unique, poetic way to add a touch of nature to your home.

Dried flowers

Dried flowers are back in style! Everlasting bouquets are an easy
way to preserve summer colors indoors. You can either try your hand
at drying them yourself or get some from your favorite florist.

IDEA 1: THE BELL
Turn terra-cotta planters into a hanging
decoration. Simply insert a piece of string
through the drainage hole and tie a big knot
to keep it from slipping. Repeat several times
with additional pots, finishing with a small
bouquet of dried flowers hanging from inside
the lowest pot.

IDEA 2: THE WALL HANGING
Tiny bouquets tied to a wooden rod with
invisible nylon thread can make a simple
but romantic wall decoration.

DIY Dried-flower wreaths

Dried flowers are perfect for making wreaths.
We've put together a few ideas to get you started.
Now it's your turn...

eucalyptus

thistle

hare's tail (bunny grass)

baby's breath

TIP If you're making a wreath that won't be entirely covered with flowers or leaves, consider wrapping the wire structure in raffia or string to conceal it.

immortelle flower

Supplies
- dried flowers
- wire-core raffia or very fine wire (that you can cut with scissors)
- metal ring (from a lampshade, or available in craft stores)
- small glue-gun

1. Using bits of raffia or wire, attach the flowers to the metal ring, moving gradually around the circumference.

2. To add small or fragile flowers with short stems, use a glue gun.

eucalyptus

love-in-a-mist seedpods

oats

rattle grass

woollyheads

sunray

DIY Foliage wreath

It's no wonder wreaths of flowers and leaves are
in style—they're beautiful and easy to make.

A few tips before you begin

It's essential to choose the right varieties of plant, because they don't all age well. All varieties of eucalyptus, myrtle, laurel, and olive are great for making long-lasting wreaths. Have fun, try everything, and feel free to add dried flowers here and there.

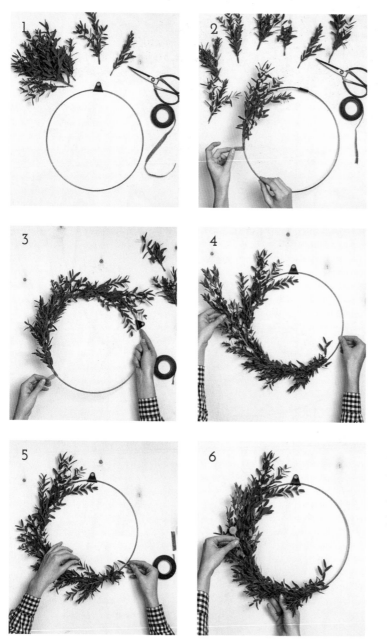

Supplies

- green leaves and dried flowers
- florist's tape or wire
- metal ring, embroidery hoop, or coat hanger, if you're working with very thick leaves.

To make large wreaths, you can also use very fine wire or wire-core raffia.

1. Assemble your equipment, trying out various combinations of dried flowers and leaves before you begin.

2. Gather several branches and fasten them together using tape.

3. to 5. Repeat the first step several times. Using tape, attach the sprigs to the ring, lightly overlapping them to hide the pieces of tape.

6. Once the wreath is finished, add dried flowers by inserting them between the leaves here and there.

TIP A wreath doesn't need to be complete to be beautiful.

DIY Leaf-trimmed hoop

Create this verdant wall decoration using delicate netting.
The flowers and leaves will come together as if by magic.

eucalyptus

myrtle

Supplies
- stems with leaves
- square of fine tulle
  or netting fabric
- embroidery hoop
- needle
- spool of thread
- pair of shears

METHOD 1
1. Mount the tulle on the embroidery hoop. Insert the stem through the netting, front to back.

2. to 3. Slip the stem below the netting, then bend it back through the netting, back to front, to fix it in place.

METHOD 2
1. Mount the tulle on the embroidery hoop. Insert the stem through the netting, front to back.

2. to 3. Slip the stem below the tulle and position it to your liking.

4. Using your needle and thread, wrap the stem a little higher up and pass the thread under the tulle, as if you were sewing the stem to the fabric. Tie a knot.

5. Continue adding pieces of foliage until you're happy with the effect.

6. Remember to leave some of the tulle empty. Once the piece is finished, it will look as though the foliage is suspended in mid-air.

Touches of green

Plants add poetry to our daily lives. But you don't necessarily need to buy a whole greenhouse of plants to reap the benefits. Adding just a touch of green here and there is enough.

The prettiest gift wrap: kraft paper, string, a bit of foliage, and you're done.

The perfect detail: a sprig of flowers tucked into napkins.

A bright idea to add a touch of "hygge" to
your home: candle holders made from bottles
filled with water and a few branches.

Planters

 **DIY** # Upcycled pallet garden

Nothing is wasted; everything can be transformed. Give wooden pallets a new lease of life as shelves for plants.

Supplies
- 1 pallet
- nails
- hammer
- 3 boards the same width as the pallet
- reusable bags (optional)
- heavy-duty stapler (optional)
- clay pebbles (optional)
- soil (optional)

1. If you want to paint the pallet, sand it first.

2. Hammer flat any nails sticking out of your pallet and, if necessary, reinforce the joints.

3. Turn the pallet upside-down.

4. Position three boards on one side of the crossbeams to create shelves and attach them with a few nails. And you're done.

5. You could also create flowerboxes by placing reusable bags within each new shelf. Staple the bags to the wood to hold them in place and pierce to facilitate drainage.

6. Place a layer of clay pebbles in the bottom of the flowerboxes, then soil, and last of all your plants.

 # Standing planter

Are you a fan of patterned floor tiles? Then this DIY project is for you. Watch out, though: there's a good chance you'll want to redo every surface in your house.

Supplies

- protective gloves and goggles
- 1 pine board, 7 × 8 × ¾ in. (18 × 20 × 1.8 cm)
- jigsaw
- 4 aluminum corner irons (brass-tinted), each ½ × 9½ in. (1.5 × 24.5 cm)
- metal saw
- file
- 4 ceramic tiles, each 8 × 8 in. (20 × 20 cm)
- right-angle ruler
- multi-purpose glue
- vise
- 1 terra-cotta planter, 6 in. (15 cm) tall and 6½ in. (15.3 cm) in diameter
- 1 terra-cotta saucer 6½ in. (15.3 cm) in diameter

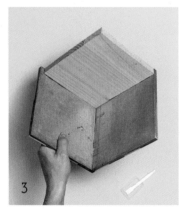

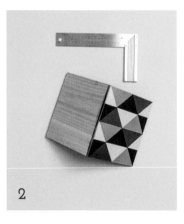

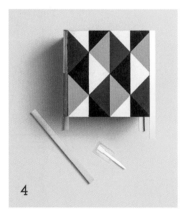

1. Wearing protective gloves and goggles, cut the pine board to size, if necessary, using the jigsaw, and adjust the corner irons using the metal saw and the file.

2. Glue one of the ceramic tiles to the pine board at a right angle, checking with the right-angle ruler and using the multi-purpose glue. Hold in place with the vise.

3. Repeat step 2 with the other tiles, gluing each side to the next.

4. Glue the four corner irons to the tiles to create legs for the planter.

DIY Hanging planter

Designer Pierre Lota had this brilliant idea
for a crafty piece with plant appeal.

1

2

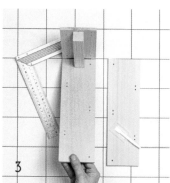

3

4

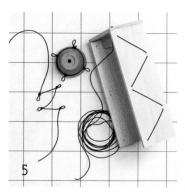

5

Supplies

- protective gloves and goggles
- carpenter's square
- pencil
- 2 pine boards, $3^3/4 \times 12 \times 1/2$ in. ($9.7 \times 30 \times 0.9$ cm)
- 1 pine board, $3^3/4 \times 6^1/4 \times 1/2$ in. ($9.7 \times 16 \times 0.9$ cm)
- drill
- wood drill bit, 0.1181 in. (3 mm) diameter
- multi-purpose glue
- 1 pine batten, $1^1/4 \times 5^3/4 \times 1^1/4$ in. ($3 \times 14.2 \times 3$ cm)
- 1 terra-cotta pot, 3 in. (8.5 cm) tall and $3^1/2$ in. (9 cm) in diameter
- 1 length polypropylene string, 10 ft. (3 m) long and $1/16$ in. (1.2 mm) in diameter

1. Mark and drill eight holes in each of the longer boards, starting with one hole in the top and bottom left corners about 1/2 in. (1.5 cm) from the edge. Trace a diagonal line from the bottom hole, drilling a second hole about $3^1/2$ in. (9 cm) away from it and about 1/2 in. (1.5 cm) from the opposite edge. Drill a third hole 1/2 in. (1.5 cm) next to it, then a fourth on the diagonal, and so on. Mark and drill six holes in the shorter board, starting with one hole in the upper right corner 1/4 in. (1 cm) from the top edge and 1/2 in. (1.5 cm) from the right-hand edge. Mark and drill a second hole 1/4 in. (1 cm) below the first. Following a horizontal line, mark and drill a third hole 1/2 in. (1.5 cm) from the

left-hand edge, as in the photo. Do the same at the other end of the board.

2. Glue the batten to the shorter board.

3. Glue this piece at a right angle to one of the longer boards.

4. Glue on the second long board, using a carpenter's square.

5. Tie four regularly spaced loop knots in a piece of string and fasten it around the pot. Pass a long piece of string through the holes in one side of the wooden planter, the knots around the pot, and the other side of the planter. Tie the string at the base to hold it in place.

Trellis of triangles

Give a climbing plant a leg up and turn your home into an urban jungle.

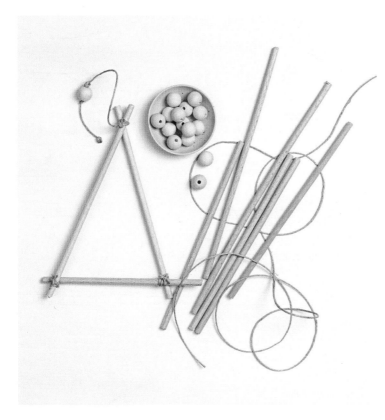

Supplies

- 6 wooden dowels, 10 in. (25 cm)
- 6 wooden dowels, 13¾ in. (35 cm)
- jute twine
- wooden beads

1. Using three dowels of equal length, make a triangle.

2. Secure two corners of the triangle with jute twine.

3. For the third and final angle, cut a longer length of twine (you'll use this twine to hang your triangle up).

4. Tie the remaining angle. Leave a certain length bare, then make another knot. Slide several wooden beads onto the string and tie a knot above to keep them in place.

5. Repeat steps 1 through 4 to make three more triangles, then hang them on the wall and watch your plant climb.

 Himmeli

Nope, this isn't a brainteaser: it's a traditional DIY project from Scandinavia. Originally made with plant stalks (rye or reeds), here it's in metal and can be used for displaying air plants.

Supplies
- 12 hollow metal tubes
- white sewing thread
- sewing needle

1. Arrange the tubes as shown in the diagram. Following the arrows, slide your thread through the tubes using the needle, then tie a knot at each end of the thread. Repeat this step on the second set of tubes, this time using a longer thread.

2. Using the thread from the second set of tubes, join the two sets of tubes together by passing the thread back through the first set. Pass the thread around the tip, then around the bottom point, and then thread it back in the opposite direction.

3. Take your time to position each tube correctly so that the structure looks like the one in the photo.

4. Position the structure on your worktop as in the photo, so that, looking from above, you have a central square and two identical triangles on each side. Bring two triangles up to meet in the center and pass your thread through the metal tubes to the top of the triangle, to form a pyramid shape. Continue to pass the thread through to secure.

5. Pick up your himmeli and thread your needle through the two lower triangles to secure them in place as before.

Air plants (*Tillandsia*) are crazy things that don't need soil; just spray them with a little water now and then. Isn't nature amazing?

DIY Hanging bottle planters

An easy way to bring a touch of the outdoors inside.

Supplies

- 6 small bottles
- jute twine
- 2 wooden dowels, 24 in. (60 cm)

1. Cut two lengths of jute twine 55 in. (140 cm) long. Arrange the dowels horizontally and parallel to one another, about 24 in. (60 cm) apart.

2. Fold one of the lengths of jute in half and wrap it once around the left end of the lower dowel. Pull the string toward the top dowel. Wrap it around the dowel and knot it tightly. Do the same thing on the right side.

3. Cut six lengths of jute twine of varying lengths.

4. Wrap a length of twine around the neck of a bottle, secure, and tie the ends of the string into a knot. You should end up with a large loop. Slip this onto the dowel.

5. Repeat for each bottle.

6. Cut a piece of twine 24 in. (60 cm) long. Knot it to each end of the top dowel and use it to hang up your structure.

This is a perfect way to display
mini bouquets and even your plant
cuttings (see page 28).

Macramé

Basic macramé knots

No book about green living would be complete without a macramé lesson. This 1970s icon is back in fashion. Once you learn these knotting techniques, you'll have plants dangling all over the place. As for color, the choice is yours.

Supplies
- 8 cords, each 6½ ft. (2 m) long
- 1 cord, 3 ft. (1 m) long

BASIC LOOP

1. Fold each of the 6½ ft. (2 m) lengths of cord in half.

2. Using the 3 ft. (1 m) cord, tie a knot around the other strings to form a loop and wrap the 3 ft. (1 m) cord around it several times.

3. Thread the cord through the loop from behind to form a loop on the right.

4. Thread the same cord through this new loop and pull it tight to form a knot.

5. Repeat steps 3 and 4 around the entire loop.

6. You will end up with a sturdy loop for your hanging planter.

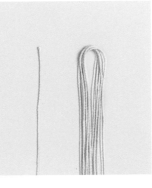

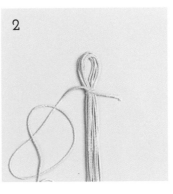

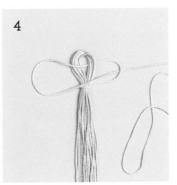

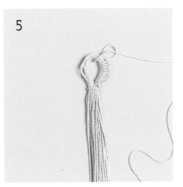

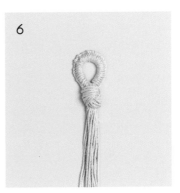

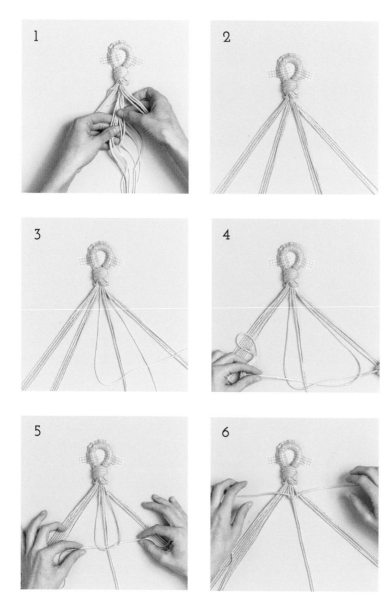

SQUARE KNOT

1. To make things easier, secure your project with tape. Separate your pieces of cord.

2. You should have four groups of four cords each, but you can also do this with just four cords.

3. Proceed by group. Take the first group of four cords. Make a half loop by bringing the cord furthest on the left over the cord in the middle and under the cord furthest right.

4. Now bring the cord furthest right under the middle cords and through the half loop formed by the cord furthest left.

5. Take hold of these two outer cords.

6. Pull the cords to make a knot. Repeat the process, but this time start by passing the cord furthest right over the cords in the middle to create a half loop, then bring the cord furthest left under the cords in the middle and through the half loop created by the cord furthest right. Repeat steps 3 through 6 until you reach the desired length.

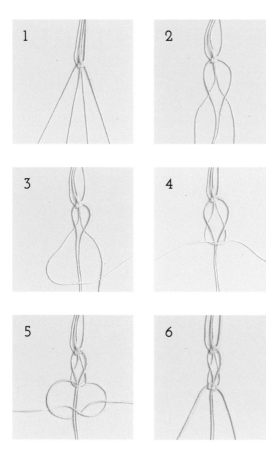

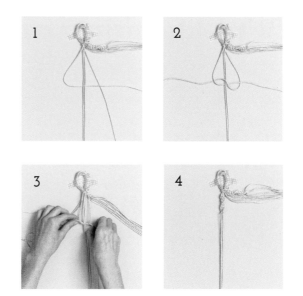

SQUARE KNOT WITH CRISSCROSS

1. Make a square knot and separate your four cords.

2. First, bring the cord furthest left over the cord immediately right. Bring the cord furthest right over the cord immediately left.

3. Join the two cords in the center and make a square knot about 2 in. (5 cm) down (see opposite).

4. You should have your first crisscross.

5. Tie two square knots.

6. Repeat steps 3 to 5.

SPIRAL KNOT (RIGHT)

1. Separate your four cords in the following way: one cord on the right, two cords in the center, one cord on the left. Make a half loop with the cord furthest left by bringing it over the two central cords and under the cord furthest right.

2. Then, bring the cord furthest right under the central strings and through the half loop formed by the cord furthest left.

3. Tighten by pulling on both cords.

4. Repeat these steps several times, always making the square knot in the same direction, to create the spiral.

SPIRAL KNOT (LEFT)

Follow the steps for the right spiral knot, reversing the direction of the square knot.

DIY Hanging planter for beginners

1

2

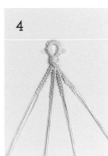
3

4

5

6

7

8

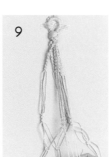
9

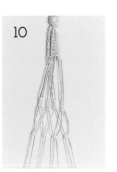
10

11

12

Supplies
- 8 cords, 6½ ft. (2 m) long
- 1 cord, 3 ft. (1 m) long
- scissors

1. to 3. Make a basic loop (see instructions on page 95), then make ten square knots with the first group of cords (see explanation on page 96).

4. Repeat steps 1 to 3 on the three other groups of cords.

5. Leave about 10 in. (25 cm) before making three more square knots.

6. Repeat step 5 on the other three groups of cords.

7. Separate each group of cords in two to obtain four Vs. Using the innermost two cords from the group furthest left and the outermost two cords from the group next to it, make three square knots.

8. Repeat step 7 with each of the remaining groups of cords.

9. and 10. Use the two outermost cords on each side to close the planter.

11. Tie all the cords in a simple knot, 4–6 in. (10–15 cm) below the last square knot.

12. Even out the cords by cutting off any excess length.

TIP Adjust the space between the knots according to the size of your pot. Doing so will enable you to make hangers for pots of different sizes.

DIY Advanced hanging planter

See instructions for basic knots on pages 95–98.

HANGING PLANTER 1

Supplies
- 8 cords, 6½ ft. (2 m) long
- 1 cord, 3 ft. (1 m) long
- scissors

1. Make a basic loop (see instructions on page 95).

2. Make fifteen spiral knots (left) (see instructions on page 97).

3. Separate the cords into two groups to form a V between them. Leave 2 in. (5 cm) of cord and tie a square knot.

4. Do this twice.

5. Make a crisscross and leave 6 in. (15 cm) before making ten spiral knots (right).

6. Separate the cords into two groups to form a V between them.

7. Leave 4 in. (10 cm) of cord bare and tie a square knot.

8. Repeat step 7, leaving 2 in. (5 cm) of cord bare.

9. Tie all the cords in a simple knot.

10. Even out the cords by cutting off any excess length.

HANGING PLANTER 2

Supplies
- 8 cords, 6½ ft. (2 m) long
- 1 cord, 3 ft. (1 m) long
- scissors

1. Make a basic loop (see instructions on page 95).

2. Tie six square knots with a crisscross between each knot. Make sure to leave the same distance between each knot so the planter is stable.

3. Repeat step 2 three times: separate the cords into two groups to form a V between them.

4. Leave 4 in. (10 cm) of cord and make a square knot. (You can also adapt the distance between the knots depending on the size of your pot.)

5. Tie all the cords in a simple knot and cut excess length as desired.

Macramé planter stand

Combine your love of plants with macramé in this pretty wooden structure for hanging planters. It's a great way to green up your home.

Supplies

- 4 wooden battens, 4 ft. × 1 in. (1.2 m × 2.7 cm)
- 1 wooden batten, 32 in. × 1 in. (0.8 m × 2.7 cm)
- 6 screws and a screwdriver

1. Screw one of the 4 ft. (1.2 m) battens to the short batten 1½ in. (4 cm) from the end—but don't tighten the screw completely.

2. Do the same at the other end of the short batten, on the same side.

3. Screw the two remaining long battens to the other side of the short batten, 2¾ in. (7 cm) from the end. Again, don't tighten the screws completely.

4. Stand the four long battens on end.

5. Cross the long battens.

6. Screw the two pairs of long battens together where they intersect.

7. Tighten the first screws.

Now brush up on your macramé (see pages 94–100)!

I love ... these old locker-room clothes
hangers transformed into plant holders.

Creative containers

These easy decoration ideas are great for creating an inspiring and plant-filled home. We've included a few twists to add even more imagination to your daily life.

Tired of flower pots? Get rid of them. Turn old shopping bags into planters for a homemade look.

Give empty jam jars new life as succulent planters.

Free the birds ... and use their cage to frame a bouquet or house a trailing plant.

This tiny cactus tucked into a spool of twine is too cute.

Try this unusual combination of air plants (*Tillandsia*) and sea urchin shells for a funky marine look. It's a great idea for air plants, which don't need soil or water (a few pumps of the spray bottle will do).

Tea time! Turn tea or coffee cups
into pots for succulents.

Make a stylish pot holder
out of a gold tin can.

Plant tiny houseplants into tea pots and canisters
(remember to drill holes in the bottom for drainage).

inspiration Fun with foliage

It's entirely possible to have a healthy respect for nature without taking yourself too seriously. Try this humorous approach to decorating with plants.

Washi tape, stickers, markers ... Anything goes when it comes to having fun. Adding a polka dot or two with a marker won't hurt your plant, so feel free to draw on the leaves. Try decorating the pots, too.

Adding a few stickers won't do anything worse than slow photosynthesis a little, so don't wait until Halloween to dress up your plants.

"Fantasy is an
eternal spring."
—Johann Friedrich von Schiller

To cure your fear of overwatering

TIP There are some plants that live entirely submerged in water, and others whose roots grow in water (ideal for flat-shaped vases).

Aquatic plants

Here's a great idea: get some aquarium plants and submerge them under water. Use stones to weigh them down in the vase. You can even add a goldfish—just remember to change the water from time to time.

P.S. A thought for our animal friends: goldfish hate round fish bowls, so try to find a large square or rectangular vase.

3. Going Green without Chlorophyll

A Green Home

Green in all its forms—lime, mint, apple, moss, olive, khaki—reminds us of the chlorophyll produced by plants and has a calming effect on our minds. So let's take a deep breath and repaint our walls.

Green evokes nature, and it seems that, the more urban our lives become, the more we dream of finding the natural world in our homes. So here are a few ideas for cultivating nature indoors.

Feng Shui masters advise against using green in the bedroom. But who really wants plants next to the bed? Besides, we have the rest of the home to decorate.

Used in the hallway, green will put your guests at ease and make them want to enter.

Green has been found to increase the appetite, so it's appropriate for the kitchen as well.

Because of its association with plant life, green gives you the impression of being in a conservatory or greenhouse, especially if it appears alongside white in a light-filled room.

Green is refreshing and bestows a feeling of comfort: you might use it in a living room, where it will encourage conversation and reduce tension.

A relaxing, soothing color that improves concentration, green should be present in every office.

Play around with the many nuances and variations of green.

A few tips for creating a harmonious, colorful room

1. Skim through magazines and cut out images of places that resonate with you. This will allow you to look more closely at the proportions of different colors used in a given space.

2. Always start with a color you really like. Ask yourself if this color works best as:
- a background color, suitable for a wall, a curtain, a bedspread—in other words, a surface several square feet in size
- an accent color, good for pillows, a small piece of furniture or a decorative object— in other words, for details

3. Add the materials found in your furniture: wood, stone, metal, fabric.

4. Add a second color, asking yourself the same question: is it a background color or an accent color? Make sure this color goes well with the first.

5. Continue in this way with four or five colors (two or three accent colors, two or three background colors).

Don't forget about quiet colors such as whites, grays, or neutrals, which will contribute to a feeling of space and calm when used on the ceiling, one or more walls, or even a large piece of furniture.

Sophie Hélène, colorist

Plant-based dyes

When it comes to plant-based dyes, the only limit is your imagination. You could spend a lifetime creating a whole palette of colors. If you use different plants and fabrics, you'll never end up with the same color twice. A world of color is waiting for you to discover.

Golden rule

Plant dyes often involve the use of a mordant. Sounds scary, right? Don't panic—it's just a technical term for a substance that ensures colors adhere to natural fibers. You can also try dying silk squares or wool muslin, since colors adhere more easily to these animal fibers, making fixatives unnecessary.

This chapter is a simple introduction to plant-based based dyes that I hope will inspire you to explore further on your own.

VERBENA DYE

Supplies
- 3 or 4 full branches of fresh verbena
- 12 cups (3 liters) rainwater (bottled water will work too)
- squares of wool muslin or silk*
- stewpot

1. Put the water and the verbena in the stewpot and bring to a boil.

2. Lower the heat and simmer for an hour.

3. Remove from the heat. Let the mixture cool overnight, then filter through a sieve to remove any debris. Put it back in the pot.

4. Dampen your fabric squares before submerging them in the dye bath.

5. Heat the pot, stopping just before the liquid comes to a boil.

6. Simmer for 1 hour, stirring from time to time.

7. Remove from the heat and let it steep overnight. Wring out the fabric and hang up to dry out of direct sunlight.

* For the dye to take, the fabric must be pre-washed to remove any existing substances.

Romarin A
Ferapres

Romarin Nov.
N.M

Romarin Nov.
N.M

Romarin Nov.
N.M

vieux drap
toile métis

Bourrette
de soie

Etamine
de laine

soie
sauvage

ROSEMARY DYE

Supplies
- several branches of fresh rosemary
- 12 cups (3 liters) rainwater (bottled water will also work)
- squares of wool muslin or silk*
- stewpot

1. Put the water and the rosemary in the stewpot and bring to a boil.

2. Lower the heat and simmer for an hour.

3. Remove from heat. Let the mixture cool overnight, then filter through a sieve to remove any debris.

4. Dampen your fabric squares before submerging them in the dye bath.

5. Heat the pot, stopping just before the liquid comes to a boil.

6. Simmer for 1 hour, stirring from time to time.

7. Remove from the heat and let it steep overnight. Wring out the fabric and hang up to dry out of direct sunlight.

* For the dye to take, the fabric must be pre-washed to remove any existing substances.

Botanical illustrations

Turn your home into a greenhouse without having
to look after a single plant.

Supplies

- engravings
- pages from botanical
  books
- double-sided tape

Using a little double-sided tape, cover your
whole wall with botanical illustrations as
though they were squares of wallpaper. Look
for engravings at flea markets, or cut out pages
from reprints of old herbals. Simple but beautiful!

DIY Embroidery for beginners

These days, embroidery is as fashionable as knitting. Forget Grandma's doilies: we're talking beautiful embroidered designs to decorate your living-room walls. Ready, set, thread your needles!

Supplies

- an existing botanical design
  (or your own design drawn on graph paper)
- tape
- pencil
- white fabric
- tracing paper
- embroidery hoop
- embroidery needle
- a nice range of green thread
- scissors

How to transfer a design to fabric

1. Take your chosen design (simple, linear designs are best).

2. If the fabric is thin and light, transfer the design by taping it to a window, then taping the fabric over the top and tracing directly on the surface.

3. If working with dark or thick fabric, start by tracing the design onto tracing paper using a sharp pencil.

4. Lay the tracing paper on the fabric, design-side down, and use the pencil to trace over the lines once more. This will transfer the pencil lines to the fabric in mirror image.

THE DRAWINGS

The designs used on the facing page are shown here in reverse, which will make them easier to transfer.

Hoop 1
Back stitch (see page 129).

Hoop 2
Back stitch used for the leaf, satin stitch used for the wreath (see page 129).

Advanced embroidery

Supplies

- an existing botanical design
  (or your own design drawn
  on graph paper)
- tape
- pencil
- white fabric
- tracing paper
- embroidery hoop
- embroidery needle
- a nice range of green thread
- scissors

Once you've mastered the basic stitches, prepare your design as before (see page 124) and try these more advanced techniques.

BACK STITCH

Make a simple, straight stitch. Bring the needle up through the fabric in front of the first stitch, where you would like the second stitch to begin, then bring the needle back down through the fabric as close as possible to the end of the first stitch. Continue in the same way, making sure the stitches are of even length.

SATIN STITCH

Bring the needle up through the fabric at your starting point, cross your design, and insert it on the opposite side of the motif you are filling. Bring your needle back up through the side you started on, very close to the first hole. Proceed in this manner, moving along the motif and keeping the stitches very close together in order to fill in the whole shape. Don't pull too tight or the fabric will pucker.

FRENCH KNOT

Bring the needle fully up through the fabric. Hold the needle horizontally, wrap the thread around the needle once (or twice for a larger knot), and reinsert it into the fabric, very close to the original hole. Pull gently to tighten the knot.

LEAF STITCH

Stitch from the tip of the leaf to the base using satin-stitch technique, alternating once to the left and once to the right, which will create the central vein.

THE DRAWINGS

The designs used on the facing page are shown here in reverse, which will make them easier to transfer.

Hoop 1

Back stitch used for stems and satin stitch used for leaves (see page 129).

Hoop 2

Back stitch used for stems, satin stitch used for small leaves, leaf stitch used for large leaves, and French knots used for buds and berries (see page 129).

Paper jungle

This is a great idea for those with a jungle spirit but no green thumb. Decorate your wall with a paper forest.

TIP Look to nature for inspiration when you create your urban jungle, or search for simple shapes online. You might even try using a variety of colors, like Matisse.

Supplies
- pencil
- large sheets of thick, colored paper (about 140 lb./300 gsm)
- scissors or box cutter
- washi tape

Trace the shape of a leaf you like onto good-quality paper or draw a large design freehand.

Cut it out.

Fix several onto your wall with washi tape.

4. Plants for Well-Being

Essential Oils and Beauty DIY
What you need to know

Essential oils are extracted from plants. They've been used for millennia for their many virtues, but we are sometimes reluctant to use them out of fear of hurting ourselves or others.

Here's some basic advice to avoid unpleasant surprises.

To start you off gently—and avoid crowding your cupboards or overwhelming you with information—we've chosen several oils that support physical and mental health, for a 100-percent natural medicine cabinet.

The following is based on advice from Sarah Canonica, a Paris-based expert in natural remedies.

Follow these basic rules

Rule No. 1
Use high-quality essential oil that is 100 percent pure, natural, unaltered, and cultivated organically by small producers. You won't do yourself any good by using bad products.

Rule No. 2
To avoid allergic reactions, do a skin patch test by placing a drop of diluted oil on your inner elbow. If nothing happens after five minutes, you can use the oil safely. If you have a reaction, clean the area with a neutral oil (almond, sunflower seed, or canola) using a cotton swab.

Rule No. 3: The golden rule
Always dilute essential oils. If they are safe to ingest, add a little to honey, olive oil, or another vegetable oil. To apply them to the body, always mix them with a carrier oil such as almond or apricot oil. Each person may have a different and unique reaction to essential oils. Remember, a higher dose doesn't necessarily mean a stronger effect. And just because an oil is high quality, that doesn't mean it will last for years.

Using essential oils safely
Never let undiluted essential oils touch your skin or be ingested. Keep them away from and out of the reach of children. If they are accidentally ingested or if they come into contact with your eyes or other mucus membranes, seek medical advice immediately. Children under seven and pregnant or breastfeeding women should not use essential oils without seeking medical supervision. If you have any existing health issues or ongoing treatments, ask your doctor before using essential oils.

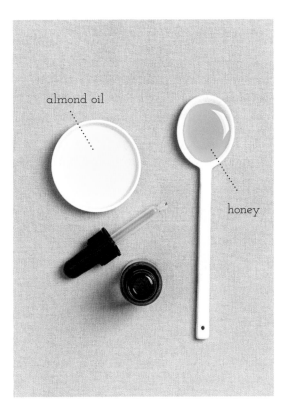

almond oil

honey

There are hundreds of essential oils, all with different properties. We've selected a dozen, divided into two categories, that can benefit mental and physical health (as you know, these two usually go hand-in-hand). After reading this introduction, you'll be fully equipped to explore further according to your own needs.

TIP Storing essential oils away from light and heat will help them stay fresh and effective for longer.

Must-have oils

When using essential oils on the body, always dilute them in almond or apricot oil. If you intend to ingest them, mix them with honey, olive oil, or another vegetable oil.

Ylang ylang Cananga odorata

English lavender Lavandula angustifolia

Cypress Cupressus sempervirens

Bay laurel Laurus nobilis

Clary sage Salvia sclarea

Five essential oils for emotional well-being

YLANG YLANG
Cananga odorata
To promote calm and cheerfulness
The "flower of flowers" encourages feelings of lightness, joy, and sensuality. Perfect for panic attacks and times of excessive stress. It can be diffused or mixed with a carrier oil for massages. Dilute a couple of drops in almond oil and massage your hands, wrists, and/or solar plexus (did you know that the little dimple under your chest is believed to be the seat of your emotions?). To create a calming atmosphere at home, use two drops on a decorative diffuser (stones, pine cones, or reeds: see pages 144–45).

CYPRESS
Cupressus sempervirens
For grounding and concentration
This oil helps us stand tall while keeping our feet solidly rooted, like a tree. It's perfect for staying focused. The night before an important meeting, mix a couple of drops with a teaspoon of almond oil and massage your feet. You'll wake up with laser-like focus.

CLARY SAGE
Salvia sclarea
Perfect for women
This oil is ideal for balancing female hormones and calming the ups and downs associated with a woman's cycle. It also stimulates the liver. Take one drop in a small spoonful of honey twice a day for a week. If you're feeling overwhelmed by your emotions, massage your solar plexus with one or two drops added to a little almond oil.

ENGLISH LAVENDER
Lavandula angustifolia
A must-have
If you buy only one essential oil, make it this one. True lavender is a fundamental material in aromatherapy. It's appropriate for everyone, children and adults alike. A sedating and soothing oil, it's perfect for improving sleep and managing stress. For a peaceful night's sleep, massage the solar plexus with a blend of one drop of lavender to one drop almond oil. Apply one or two drops to a stone and place it next to your bed to encourage restful sleep (see page 145).

BAY LAUREL
Laurus nobilis
For self-confidence
Remember Caesar's crown? This essential oil bestows confidence and assertiveness. It's best used before an interview, an exam, or a public-speaking event. Apply a blend of one drop of bay laurel oil and a small amount of almond oil to the bottom of your feet before going to bed several days before an exam. It also soothes occasional fatigue. Take one or two drops in a teaspoon of honey, olive oil, or mixed into almond paste three times a day for seven days.

Seven essential oils for physical well-being

BASIL
Ocimum basilicum
Soothes monthly cramps
This oil is extremely effective against stomach cramps. You'll never need ibuprofen again! It's best used as a massage oil: dilute two drops in eight drops of almond oil. One or two drops taken with a teaspoon of honey can also help ease digestive troubles.

BLACK SPRUCE
Picea mariana
Invigorating
This oil helps fight winter fatigue by stimulating and strengthening the immune system. Mix two drops with four drops of almond oil and rub it on your lower back in the morning for three weeks. Never use it at night. It can also be used in a diffuser as an air purifier. Use ten drops in a capful of unscented shower gel or honey for an invigorating and fortifying bath.

KATAFRAY
Cedrelopsis grevei
For muscle aches and pains
Fights fatigue and is a powerful anti-inflammatory. Perfect after moving house or running a marathon. Dilute two drops in eight drops of almond oil. Massage the affected area three to five times daily until the pain disappears.

PEPPERMINT
Mentha piperita
Stimulating
This essential oil works wonders on occasional fatigue. It also stimulates digestion to ward off the post-meal slump. Take two drops two or three times a day in a teaspoon of honey or olive oil for no more than seven days at a time. It also soothes headaches (used alone or with lavender). Combine one drop with four drops of almond oil and massage the temples one to three times a day.

RAVINTSARA ("the good leaf tree")
Cinnamomum camphora
A winter must-have
Another incredibly useful oil. Appropriate for adults and children alike, it protects against winter infections; it's also good for treating the onset of a cold. As a preventative treatment, mix two drops with a few drops of almond oil and apply to the wrists, breathing deeply, once or twice a day. If the cold has settled in, apply three to four times a day. You can also take it with a spoonful of honey or apply to the sole of each foot before going to bed, if you have a cold. Stop using for one week after three weeks of use.

ROSEMARY
Rosmarinus officinalis L. verbenoniferum (Rosemary Verbenone, not to be confused with other varieties)
Detoxifying
Good for stimulating the liver and supporting healthy liver function. Take one drop in a teaspoon of honey or olive oil for seven days. Alternatively, mix one drop with a teaspoon of olive oil and gently massage the area twice a day.

SARO
Cinnamosma fragrans Baillon
Drives colds away
This oil complements ravintsara and will help you fight off a bad cold. Dilute two drops in two drops of almond oil. Massage onto feet before bed, or mix one drop with honey or water and take internally.

DIY Essential-oil diffusers

A simple way to create your own aromatherapy diffuser.

Supplies

- a glass jar (not plastic)
- vodka
- essential oil
- wooden skewers

Pour two parts vodka (from last night's party) into the glass jar. Add one part water. Finally, add sixty to eighty drops of essential oil, depending how strong an odor you would like and the size of the room. You can also, of course, blend oils for multiple benefits.

Trim the sharp edges from the skewers and plunge them in the liquid; they will help diffuse the odor.

When choosing your oils, listen to your body: it knows what it needs.

You can also create a specific blend for each room in your home: a peaceful scent in the bedroom to help you sleep soundly, for example, and a more energetic scent in the bathroom for starting the day on the right foot.

For a calming scent, use a combination of ylang ylang, frankincense, and sandalwood.

For an uplifting scent, try mixing sweet orange, grapefruit or lemon, and lavandin (French lavender).

For a purifying perfume, try ravintsara and niaouli (broad-leaved paperbark).

TIP You can use your blend to create a fragrant room spray by simply putting it in a spray bottle.

Place these simple, natural diffusers throughout your home,
in your closets, or give some to friends.

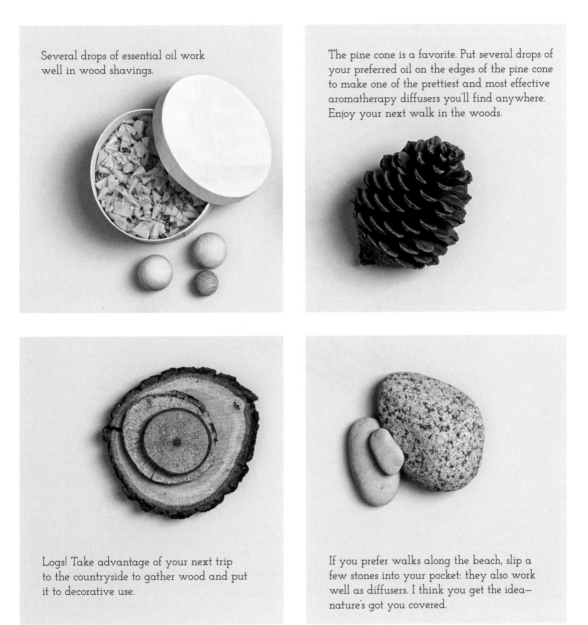

Several drops of essential oil work
well in wood shavings.

The pine cone is a favorite. Put several drops of
your preferred oil on the edges of the pine cone
to make one of the prettiest and most effective
aromatherapy diffusers you'll find anywhere.
Enjoy your next walk in the woods.

Logs! Take advantage of your next trip
to the countryside to gather wood and put
it to decorative use.

If you prefer walks along the beach, slip a
few stones into your pocket: they also work
well as diffusers. I think you get the idea—
nature's got you covered.

 # Lip balm

Are you careful about what you put on your body?
You should be! These two easy-to-make products
are a great introduction to home-made cosmetics.

Ingredients

- 2 tsp. beeswax
- 3 tbsp. shea butter or
  macadamia oil
- 2 drops lavender or rosemary
  essential oil
- small container

1. In a bain-marie, melt the beeswax and shea butter or macadamia oil.

2. Remove from the heat. Stir well and add the essential oil.

3. Pour the mixture into a small container, preferably glass, and place it in the refrigerator to harden.

BE CAREFUL
In summer, avoid applying essential oils to your face and other exposed parts of the body. The summer sun can cause photosensitivity, resulting in unsightly blotches or burns.

Hydrating body lotion

Ingredients
- 1 tbsp. coconut oil
- 1 tbsp. almond oil
- 2 tbsp. shea butter
- 4 drops lavender or ylang ylang essential oil
- air-tight glass container

1. Whisk the coconut and almond oils with the shea butter to obtain a creamy paste.

2. Add the essential oils and store in an air-tight glass container.

TIP You can use other oils of your choice. Argan and jojoba oils also penetrate easily and moisturize well.

SUPER ORGANIC TIPS
- For a good toothpaste, mix coconut oil and turmeric.
- For a simple deodorant, add a pinch of baking soda to a little coconut oil.
- For easy make-up removal, use apricot kernel or almond oil (even waterproof mascara is no match for these guys), and finish with a floral water (rose, cornflower, chamomile, orange flower, witch hazel, and rockrose are all good choices).

Infusions and Recipes
What you need to know

The following is based on advice from Véronique Cappello,
a plant therapist and aromatherapist based in Paris.

Let's not forget: plants are our friends.

They brighten up our lives and do our minds and bodies a world
of good. Plus, they have undeniable medicinal properties. Follow
these tips below and start taking care of yourself with infusions:

- Use plants that you know have been grown organically
  and buy from small producers or reputable herbalist shops.
- Check the Latin name, origin, and parts of the plant
  being used.
- Never combine more than three plants.
- Avoid long-term use unless under medical supervision.

METHOD
1. Add a tablespoon of the dried plant or blend to a teapot.
2. Pour 1 cup simmering water over the leaves and let steep for 10–15 minutes,
covered. Strain.
3. Drink morning and evening for three weeks or use occasionally, depending
on your needs.

Remember, plants have real medicinal properties. Don't mix infusions
and prescription medication without consulting your doctor first, as
they may interact.

Pregnant and breastfeeding women, children, and anyone
with health issues should seek the advice of a medical professional.

Must-have infusions*

Hundreds of plants are used in phytotherapy. The following "essentials" will become your best friends for life.

California poppy

Eschscholzia californica
(entire plant, flowering)
A very effective anti-spasmodic, sedative, and anti-anxiety plant. It helps regulate sleep by reducing the time it takes to fall asleep and by prolonging sleep. Off to bed with you! Warning: to be avoided if you suffer from glaucoma or are pregnant.

> "Let the power of plants take care of you."

Purple passionflower

Passiflora incarnata
(above-ground parts)
Anti-spasmodic, anti-anxiety, and calming, it is a gentle remedy for anxiety, tension, and hypersensitivity to sounds and odors. A gentle medicine.

Hawthorn

Crataegus monogyna and *Crataegus laevigata*
(leaves and flowers)
Helps reduce stress-related palpitations (the feeling that your heart is racing or beating hard) and nighttime stress with insomnia. Basically, it's good for anyone who's overworked and stressed out.

Thyme

Rosemary

Thymus vulgaris
(leaves and flowering tips)
This natural antiseptic is also warming, which
makes it useful when winter arrives. At the first
sign of a shiver, make yourself a cup. Special dosage:
one teaspoon, steep 5–10 minutes.

Rosmarinus officinalis
(leaves and flowering tips)
Rosemary increases the production of bile and helps
ease digestive issues by supporting the liver and
gallbladder. Stimulating and energizing, it's also good
for getting you back on your feet after an illness
or period of fatigue. Warning: do not use if you suffer
from hypertension.

Lemon balm

Yarrow

Melissa officinalis (leaves)
Calms the nervous and digestive systems. Ideal
for stress that manifests itself in the stomach.
Soothes and calms painful menstruation associated
with underlying anxiety. Warning: do not use
if you suffer from thyroid disease.

Achillea millefolium
(flowering tips)
Good for the liver and stomach, it stimulates digestion
and calms premenstrual cramps. Do not use if you
are allergic to Astèraceae (lettuce, endive,
chicory, Jerusalem artichoke, etc.).

DIY Infusions for well-being

These beneficial brews are easy to prepare. Make sure to use organic products from small producers.

RELAXING INFUSION

Ingredients

- 2 cups (500 ml) water
- 3 branches of thyme
- several leaves of lemon verbena
- several chamomile flowers
- honey (according to taste)

1. In a saucepan, heat the water.

2. Pour the simmering water over the plants.

3. Let steep, covered, for 10 minutes. Strain and drink.

DETOX INFUSION

Ingredients
- 1 lemon
- fresh ginger (about ¾–1¼ in./1.5–3 cm, depending on desired strength)
- several whole sprigs of mint
- 2 cups (500 ml) hot water

1. Cut two thin slices of lemon and squeeze the rest of the fruit into a teapot or saucepan.

2. Chop the ginger stem (use more if you like a stronger flavor).

3. Put the ginger and the mint into the pot. Add hot water and let steep for several minutes.

You can drink this infusion right away or store it in the refrigerator and drink chilled. Keeps for one day.

DIY Waters and "green juice"

They're beautiful and good for you! Try these enhanced waters and green smoothies.

TASTY DETOX WATER

Ingredients
- 4 cups (1 liter) filtered or spring water
- 4 untreated rose buds
- 1 medium-sized cucumber

Combine all the ingredients in a bottle and let steep at least 30 minutes before drinking.

REFRESHING DETOX WATER

Ingredients
- 4 cups (1 liter) filtered or spring water
- 1 lemongrass stick
- 1 lime
- 1 sprig of lemon thyme

Combine all the ingredients in a bottle and let steep at least 30 minutes before drinking. In summer, serve this water nice and fresh.

GREEN SMOOTHIE
FOR HEALTHY DIGESTION

Ingredients
- ½ ripe avocado
- a large handful of lamb's lettuce
- 6 sprigs of mint
- zest and juice of ½ lime
- ¼ glass of filtered or spring water, if necessary

Combine all the ingredients in a blender and mix to obtain a smooth liquid. If the smoothie is too thick, add water. Drink immediately to benefit from all the minerals and vitamins. Lamb's lettuce and mint are good for the digestive tract, so this smoothie will help regulate digestion.

DIY · Eat your greens!

This isn't a cookbook, but we wanted to share this healthy and delicious recipe with you.

GREEN SALAD

Ingredients

Serves two as a main dish or four as a side dish
- ¾ cup (130 g) uncooked quinoa
- a handful of beet greens
- a handful of dandelion greens
- a handful of young spinach leaves
- 6 generous tbsp. olive oil
- a bunch of fresh basil, rinsed
- a handful of beet sprouts
- a handful of pea sprouts
- ½ cup (75 g) feta
- ¼ cup (40 g) toasted pine nuts
- salt and pepper

1. Cook the quinoa in salted water at a rolling boil for 12 minutes. It's done when the grain is transparent and the white germ is visible.

2. Transfer the quinoa to a sieve and rinse it in cold water to halt cooking. Set aside in the refrigerator.

3. Rinse the greens and spinach leaves. With the olive and basil, make a quick pesto using a mortar and pestle or a small food processor.

4. Pour the sauce on the quinoa and mix well. Add salt and pepper to taste.

5. Add the greens and sprouts and mix again.

6. Sprinkle with the feta and toasted pine nuts.

Thank you!

Acknowledgments

You can't imagine how excited I was when my editor and I decided to launch this project. But I was also a little worried because I've never had a green thumb. I've always been a fan of nature's beauty without understanding any of its secrets. So, to do this book right, I surrounded myself with people who do. I can't thank them enough for their kindness, hospitality, and cooperation. Thank you, each and every one of you. This is your book, too, and together we pulled off an incredible feat, a beautiful introduction to sharing an appreciation and respect for nature.

I am very grateful to Clélia Ozier-Lafontaine, my editor, for her energy and enthusiasm for DIY projects. We connected instantly, and the project went off without a hitch (I know, that never happens, but there's a first time for everything, right?).

An enormous thank you to Florence Deviller, an incredibly talented artistic director (as well as a great friend), without whom this book would never have come to be. @lisamonagram

Thank you to Frédéric Baron-Morin, an excellent photographer who indulges my most outrageous projects with an enormous capacity for adaptation. @baronmorin

Thank you to Caroline Ciepielwski, the owner of Mama Petula and a pioneer of the green lifestyle who enjoys sharing her love for nature, for being the first to agree to take part in this adventure. Kudos for the swimming-pool clothes hangers, page 107. Find her in one of her two stores at Grands Voisins (in the Oratoire building, Paris 14th arr.) and Ground Control (Paris 12th arr.), and on her Instagram account @mamapetula

An enormous thank you to Agnès Valverde, owner of the incredibly beautiful flower and plant shop Éphémère, who, along with her wonderful team, provided much help and support throughout the book's creation. She's kind of my fairy godmother. A special thank you also to Beatrice Heuze for her marvelous crowns and wreaths, pages 70–73. Éphémère, 133 Avenue Parmentier, 75011 Paris and @ephemerefleuriste

Thank you to Camille and Pauline Laffargue from the Cactus Club for a chapter bristling with good ideas. Meet them in their lovely store, recently renovated by Atelier ADC, at 29 rue de la Fontaine au Roi, Paris 11th arr., and on their Instagram account @lecactusclub

Thank you to Noam Levy at Green Factory, the world's trailblazing terrarium shop, for his precious advice and lovely tutorials. Every day, he finds a new way to live in harmony with plants. A special thank you to Elodie Vitman for her energy and Mathilde Lacoste for the tutorials. Find Noam at one of his two Paris shops: 17 rue Lucien Sampaix, Paris 10th arr., or 98 rue des Dames, Paris 17th arr. @greenfactory

A very big thank you to phyto-aromatherapist Véronique Cappello for her priceless expertise on the power of plants and her introduction to infusions. Trained in traditional Chinese medicine and a graduate of the École des Plantes de Paris (Paris Plant School), she sees patients in Paris and Samoreau (south Seine-et-Marne region) and offers a lovely alternative to standard treatments. She also teaches at the École des Plantes de Paris and is regularly invited to other establishments and events. @veroniquecappello

A huge thank you to Sarah Canonica, who oversees an extraordinary health and well-being store and offers a large selection of high-quality essential oils made by small producers, as well as high-quality treatments. Thank you for an incredible job explaining essential oils and for the wonderful ideas for diffusers. Les Huiles, 34 bis rue Bichat, Paris 10th arr. @leshuiles_inspirées_bichat

Thank you to Marie Pourrech at Maison Bastille for her delicious recipes, pages 154–57.
Maison Bastille, 34 bis Rue Amelot, Paris 11th arr., and on Instagram @maisonbastille

Thank you to Pierre Lota, a designer as talented as he is nice, for his DIY tutorials, pages 84–87. @pierrelota

Thank you to Anne-Charlotte Cauville, a miracle worker of everything hand-made and owner of Aimeraude Atelier, for her patience and the creation of many DIY tutorials in this book: pages 74–79, 88–101, 132–35, 146, 147, 152, and 153. @aimeraudeatelier

Thank you to my favorite embroiderer, Juliette Michelet, for her talent and incredible energy, pages 124–31. @juliettemichelet

A special mention for Flore Avent, our beautiful model and a fan of tattoos, plants, and DIY projects.

Thank you to the entire team at Welcome Bio, who welcomed me warmly for photo shoots, with a special call-out to Julie Morel for her hospitality. Thank you to Rebecca Alindret for the "upcycled pallet" tutorial, page 82, and Bastianine Bideau for the beautiful photos of her and her charming "flower-pot bell," page 69. Visit their amazing boutique at 10-11-13 rue Boulle, Paris 11th arr. @welcomebio_bazar

Thank you to Laetitia Michel, owner of the beautiful No-Bla Bla, an inspiring boutique where she presents her textile collections and design universe. She also hosts DIY workshops, pages 10 and 102.
23 rue Keller, Paris 11th arr. @no_blabla

Thank you to Sophie Hélène, artist and colorist, for her expertise in color, pages 114–17. @sophieheleneartiste

Thanks to Bernadette Teyras who shared her passion for plant dyes with us from her garden in Auvergne, pages 118–21. @bernadetteteyras

SUPPLIES
Thank you to all the incredible stores that loaned me everything I needed to create beautiful images.
A big thanks to Truffaut, Fleux, Mosaic del Sur, La Cerise sur le Gâteau, Linge Particulier, Lilipinso, Papermint, Pompom Bazar, and La Trésorerie.
A special shout-out to Madeleine and Gustave, whom I cleaned out :)
A special thank you to @myoldpapers for the wonderful botanical illustrations.
Thank you @herbarium for lending the lovely fern that appears in the section on pressed leaves.

ENGLISH EDITION
Editorial Director: Kate Mascaro
Editor: Sam Wythe
Translated from the French by Kate Robinson
Design: Florence Deviller
Copyediting: Lindsay Porter
Typesetting: Gravemaker+Scott
Proofreading: Nicole Foster
Production: Christelle Lemonnier
Color Separation: IGS
Printed in Spain by Indice

Originally published in French as *Végétale thérapie*
© Flammarion, S.A., Paris, 2018

English-language edition
© Flammarion, S.A., Paris, 2019

19 20 21   3 2 1

ISBN: 978-2-08-020389-2

Legal Deposit: 03/2019

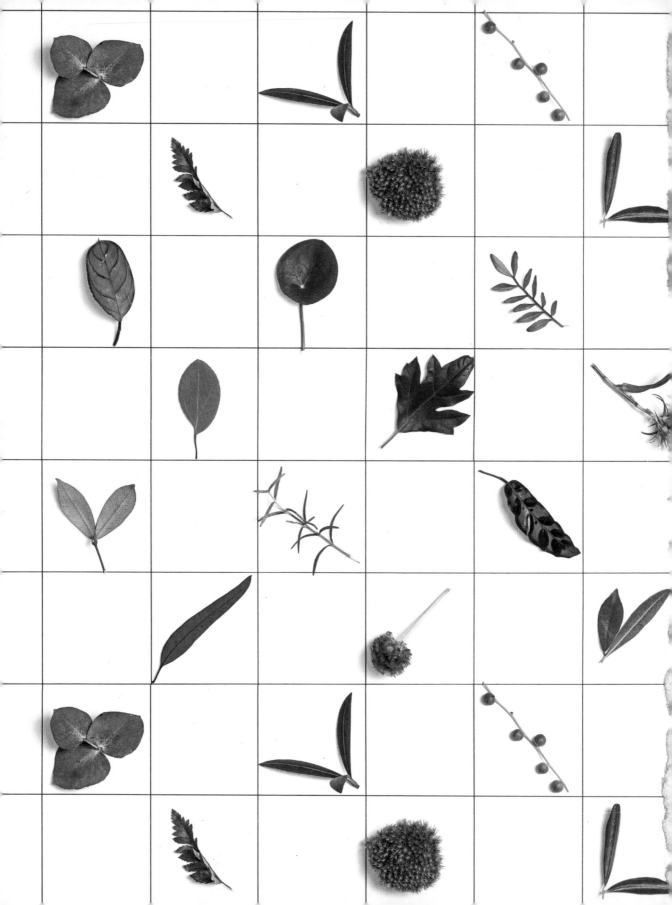